DIARY**2018**

Moon Phases

- ● New Moon
- ☽ First Quarter
- ○ Full Moon
- ☾ Last Quarter

Frances Lincoln Limited
74–77 White Lion Street
London N1 9PF
www.quartoknows.com

Science Museum Diary 2018

First Frances Lincoln edition 2017

Produced under licence for SCMG Enterprises Ltd.
Science Museum ® SCMG.
Every purchase supports the museum.
www.sciencemuseum.org.uk.

A catalogue record for this book is available from the British Library

ISBN: 978-0-7112-3856-5

Printed in China

9 8 7 6 5 4 3 2 1

Title page: Directional Robot, made by Yonezawa, Japan, for Cragstan, USA, 1957.
Below: Handcoloured lithograph of butterflies from John O. Westwood's edition of Dru Drury's *Illustrations of Exotic Entomology*, 1837.

Quarto Knows

Quarto is the authority on a wide range of topics.

Quarto educates, entertains and enriches the lives of our readers – enthusiasts and lovers of hands-on living.

www.QuartoKnows.com

CALENDAR 2018

JANUARY

M	T	W	T	F	S	S
1	2	3	4	5	6	7
8	9	10	11	12	13	14
15	16	17	18	19	20	21
22	23	24	25	26	27	28
29	30	31				

FEBRUARY

M	T	W	T	F	S	S
			1	2	3	4
5	6	7	8	9	10	11
12	13	14	15	16	17	18
19	20	21	22	23	24	25
26	27	28				

MARCH

M	T	W	T	F	S	S
			1	2	3	4
5	6	7	8	9	10	11
12	13	14	15	16	17	18
19	20	21	22	23	24	25
26	27	28	29	30	31	

APRIL

M	T	W	T	F	S	S
						1
2	3	4	5	6	7	8
9	10	11	12	13	14	15
16	17	18	19	20	21	22
23	24	25	26	27	28	29
30						

MAY

M	T	W	T	F	S	S
	1	2	3	4	5	6
7	8	9	10	11	12	13
14	15	16	17	18	19	20
21	22	23	24	25	26	27
28	29	30	31			

JUNE

M	T	W	T	F	S	S
				1	2	3
4	5	6	7	8	9	10
11	12	13	14	15	16	17
18	19	20	21	22	23	24
25	26	27	28	29	30	

JULY

M	T	W	T	F	S	S
						1
2	3	4	5	6	7	8
9	10	11	12	13	14	15
16	17	18	19	20	21	22
23	24	25	26	27	28	29
30	31					

AUGUST

M	T	W	T	F	S	S
		1	2	3	4	5
6	7	8	9	10	11	12
13	14	15	16	17	18	19
20	21	22	23	24	25	26
27	28	29	30	31		

SEPTEMBER

M	T	W	T	F	S	S
					1	2
3	4	5	6	7	8	9
10	11	12	13	14	15	16
17	18	19	20	21	22	23
24	25	26	27	28	29	30

OCTOBER

M	T	W	T	F	S	S
1	2	3	4	5	6	7
8	9	10	11	12	13	14
15	16	17	18	19	20	21
22	23	24	25	26	27	28
29	30	31				

NOVEMBER

M	T	W	T	F	S	S
			1	2	3	4
5	6	7	8	9	10	11
12	13	14	15	16	17	18
19	20	21	22	23	24	25
26	27	28	29	30		

DECEMBER

M	T	W	T	F	S	S
					1	2
3	4	5	6	7	8	9
10	11	12	13	14	15	16
17	18	19	20	21	22	23
24	25	26	27	28	29	30
31						

CALENDAR 2019

JANUARY

M	T	W	T	F	S	S
	1	2	3	4	5	6
7	8	9	10	11	12	13
14	15	16	17	18	19	20
21	22	23	24	25	26	27
28	29	30	31			

FEBRUARY

M	T	W	T	F	S	S
				1	2	3
4	5	6	7	8	9	10
11	12	13	14	15	16	17
18	19	20	21	22	23	24
25	26	27	28			

MARCH

M	T	W	T	F	S	S
				1	2	3
4	5	6	7	8	9	10
11	12	13	14	15	16	17
18	19	20	21	22	23	24
25	26	27	28	29	30	31

APRIL

M	T	W	T	F	S	S
1	2	3	4	5	6	7
8	9	10	11	12	13	14
15	16	17	18	19	20	21
22	23	24	25	26	27	28
29	30					

MAY

M	T	W	T	F	S	S
		1	2	3	4	5
6	7	8	9	10	11	12
13	14	15	16	17	18	19
20	21	22	23	24	25	26
27	28	29	30	31		

JUNE

M	T	W	T	F	S	S
					1	2
3	4	5	6	7	8	9
10	11	12	13	14	15	16
17	18	19	20	21	22	23
24	25	26	27	28	29	30

JULY

M	T	W	T	F	S	S
1	2	3	4	5	6	7
8	9	10	11	12	13	14
15	16	17	18	19	20	21
22	23	24	25	26	27	28
29	30	31				

AUGUST

M	T	W	T	F	S	S
			1	2	3	4
5	6	7	8	9	10	11
12	13	14	15	16	17	18
19	20	21	22	23	24	25
26	27	28	29	30	31	

SEPTEMBER

M	T	W	T	F	S	S
						1
2	3	4	5	6	7	8
9	10	11	12	13	14	15
16	17	18	19	20	21	22
23	24	25	26	27	28	29
30						

OCTOBER

M	T	W	T	F	S	S
	1	2	3	4	5	6
7	8	9	10	11	12	13
14	15	16	17	18	19	20
21	22	23	24	25	26	27
28	29	30	31			

NOVEMBER

M	T	W	T	F	S	S
				1	2	3
4	5	6	7	8	9	10
11	12	13	14	15	16	17
18	19	20	21	22	23	24
25	26	27	28	29	30	

DECEMBER

M	T	W	T	F	S	S
						1
2	3	4	5	6	7	8
9	10	11	12	13	14	15
16	17	18	19	20	21	22
23	24	25	26	27	28	29
30	31					

Tafel II.

Fig I.

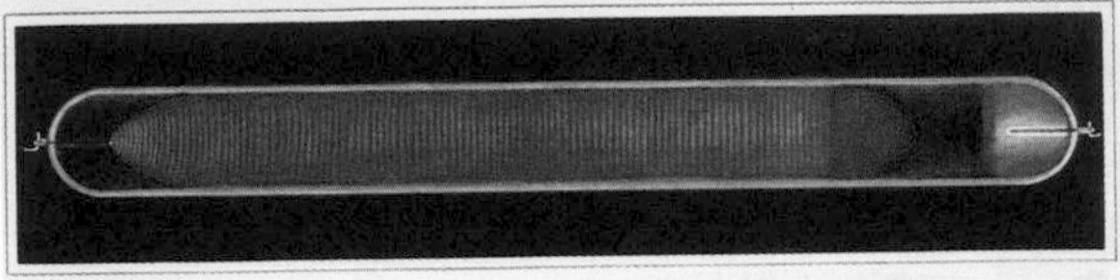

Fig II a.&b.

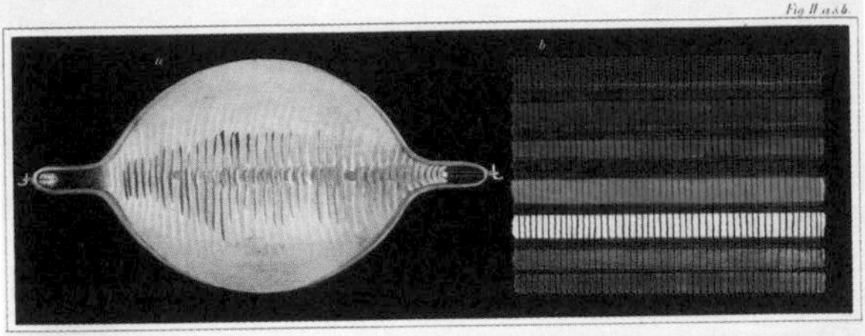

a

b

Fig III.

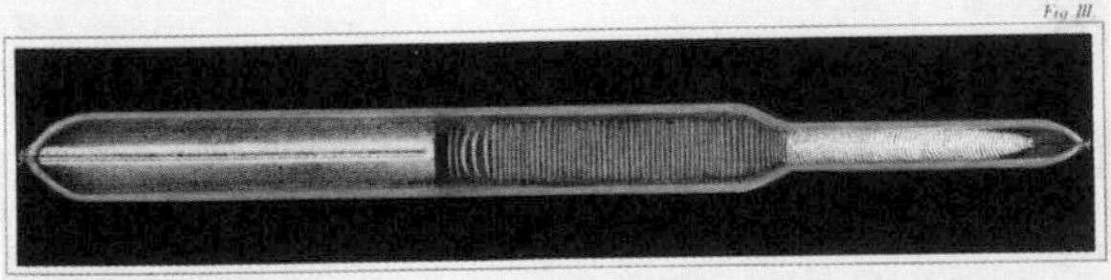

Fig IV.

INTRODUCTION

The Science Museum is the most popular destination for science, technology and engineering in the UK. Offering visitors of all ages an incredible collection of objects, both historical and cutting edge, as well as contemporary science learning and debate, we help make sense of the science that shapes our lives and gives inspiration to scientists of the future.

The museum's library collections include books, journals, patents and maps charting the worldwide development of science, engineering and medicine from the fifteenth to the twenty-first century. From these rich archives we have compiled this fascinating selection of images that show how the beauty and detail of scientific illustration and photography can have an appeal beyond its historical value.

The wide range of images include a robot dating from 1957, a close-up of one of the moons of Jupiter, a hand-coloured engraved plate illustrating the science of optics from 1850, an x-ray positive print from the First World War, an engraving of Herschel's 20-foot telescope made in 1783, a geological map of the Earth's moon and a photograph of Thomas Edison's filament lamp.

Illustration of glowing Geissler tubes, showing electric discharges through gases, from *Beobachtungen uber das geschichtete electrische Licht* (Observations on the History of Electric Light) by W.H. Theodor Meyer, published in 1858.

JANUARY

WEEK **01**

01
MONDAY

New Year's Day
Holiday, UK, Republic of Ireland, USA, Canada, Australia and New Zealand

02
TUESDAY

○ Holiday, Scotland and New Zealand

03
WEDNESDAY

04
THURSDAY

05
FRIDAY

06
SATURDAY

Epiphany

07
SUNDAY

Raketen Fahrt (Rocket Travel), a book on the possibilities of travel by rocket by Max Valier, published in 1930.

RAKETEN FAHRT
VON MAX VALIER

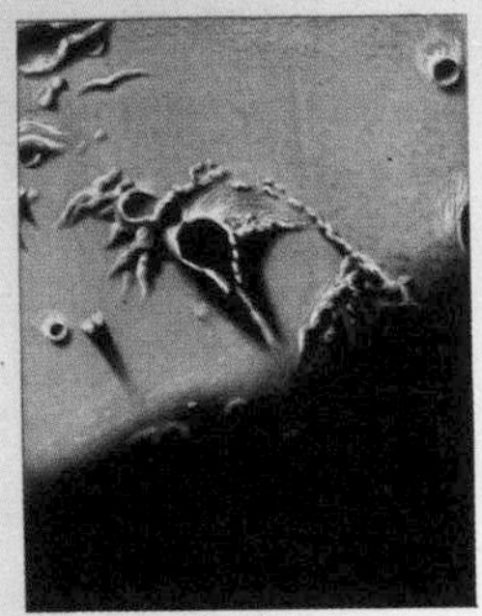

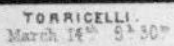

TORRICELLI.
March 14th 9h 30m

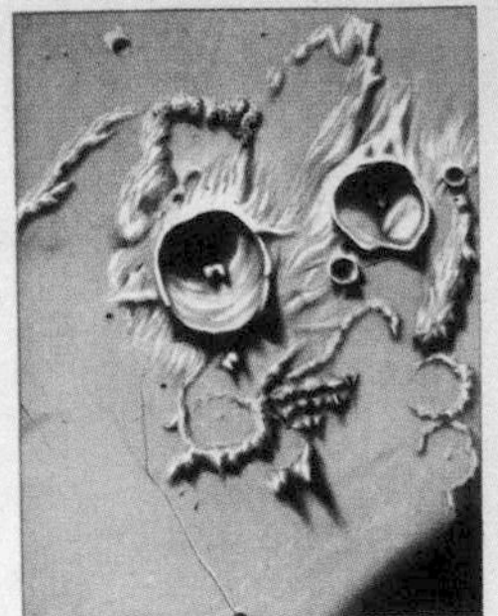

AGRIPPA & GODIN.
March 16th 8h 00m

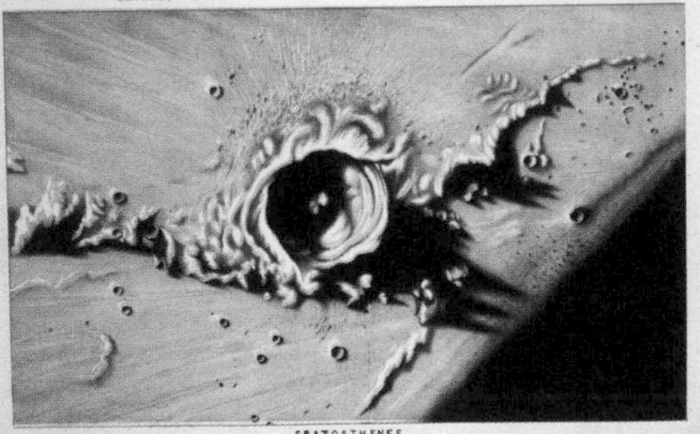

ERATOSTHENES.
April 16th 9h 00m

JANUARY

WEEK **02**

08
MONDAY

09
TUESDAY

10
WEDNESDAY

11
THURSDAY

12
FRIDAY

13
SATURDAY

14
SUNDAY

Lithographic print by Etienne Leopold Trouvelot from 1872 showing three sketches of craters on the Moon (Torricelli, Agrippa with Godin and Eratostenes), issued by Harvard College Observatory in 1876.

JANUARY

WEEK **03**

15
MONDAY

Holiday, USA (Martin Luther King Jnr Day)

16
TUESDAY

17
WEDNESDAY

18
THURSDAY

19
FRIDAY

20
SATURDAY

21
SUNDAY

Colour lithograph of the 'Eagle', an airship designed by the Compte de Lennox in 1834 to create a direct communication link between the capitals of Europe.

An exact representation
of the first Aerial Ship

THE EAGLE,

now exhibiting at the grounds of
The Aeronautical Society
Victoria Road, facing Kensington Gardens.

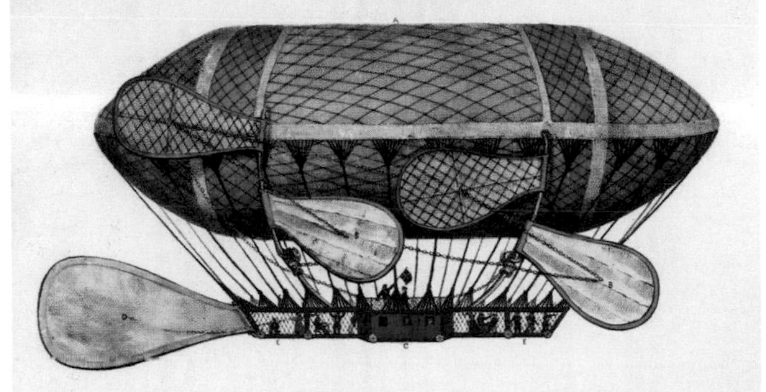

This stupendous Machine is 160 feet long 50 high and 40 wide constructed for establishing a direct communication between the Capitals of Europe, the first experiment of this new System of Aerial Navigation will be made from London to Paris and back again early in August.

- A *The Body of the Vessel in the construction of which upwards of 1400 Yards of lawn has been consumed and is capable of containing 1200 Cubic feet of Gas. There are four wings on each side marked* B *made with moveable flaps with net work on one side to support them whilst propelling the Vessel*
- C *The Cabin which contains the Machinery for working the wings*
- D *The Rudder*
- E *The extremities of the car which is made of wood, and is 75 feet long & 6 wide strongly secured on all sides by*

JANUARY

WEEK **04**

22
MONDAY

23
TUESDAY

24
WEDNESDAY

☽

25
THURSDAY

26
FRIDAY

Holiday, Australia (Australia Day)

27
SATURDAY

28
SUNDAY

Photograph of Ganymede, Jupiter's largest moon, showing the site of an impact on the icy surface which produced a crater and sprayed clean ice from below the surface over the landscape, taken by Voyager 2 in 1979.

JANUARY • FEBRUARY

WEEK **05**

29
MONDAY

30
TUESDAY

31
WEDNESDAY

○

01
THURSDAY

02
FRIDAY

03
SATURDAY

04
SUNDAY

Engraving showing a double-acting condensing rotative engine, taken from Reynolds' *Pictorial Atlas of Arts, Sciences, Manufacturers and Machinery*, published in the 19th century.

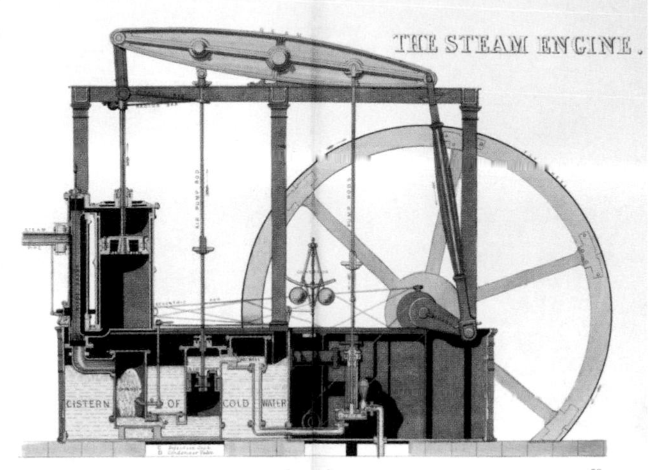
THE STEAM ENGINE.
AIR PUMP ROD
CISTERN OF COLD WATER

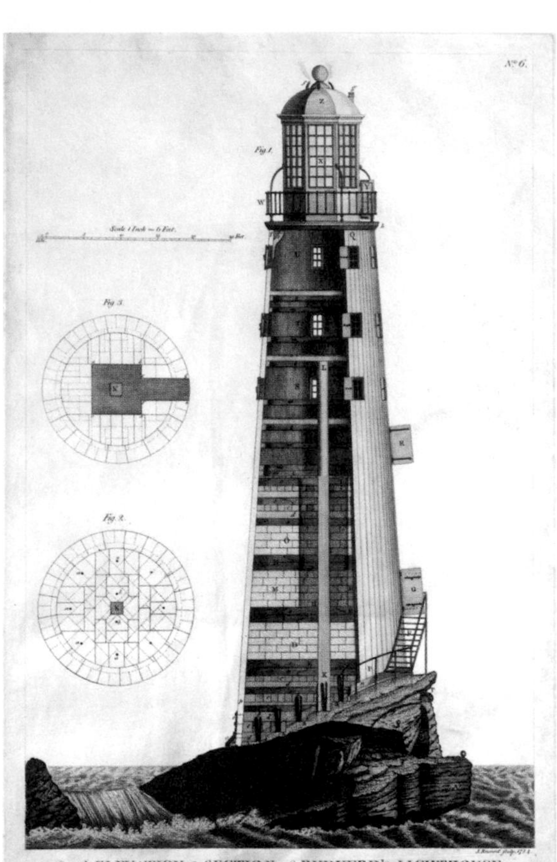

South ELEVATION & SECTION *of* RUDYERD's LIGHTHOUSE,

Compleated in 1709; represented as it stood previous to its demolition by Fire, in the Year 1755.

FEBRUARY

WEEK **06**

05
MONDAY

06
TUESDAY

Accession of Queen Elizabeth II
Holiday, New Zealand (Waitangi Day)

07
WEDNESDAY

☾

08
THURSDAY

09
FRIDAY

10
SATURDAY

11
SUNDAY

Engraving from 1784, based on a cross-sectional drawing of the second Eddystone lighthouse made in the year of its destruction in 1755. Designed by John Rudyerd and built between 1706 and 1709, this was the second of four lighthouses to be built on Eddystone Rocks.

FEBRUARY

WEEK **07**

12
MONDAY

13
TUESDAY

Shrove Tuesday

14
WEDNESDAY

Valentine's Day
Ash Wednesday

15
THURSDAY

16
FRIDAY

Chinese New Year

17
SATURDAY

18
SUNDAY

Lithograph poster of Russian cosmonaut Yuri Gagarin, produced in the Soviet Union in 1973. In 1961 Gagarin became the first man to travel in space, completing a circuit of the Earth in the Vostok spaceship.

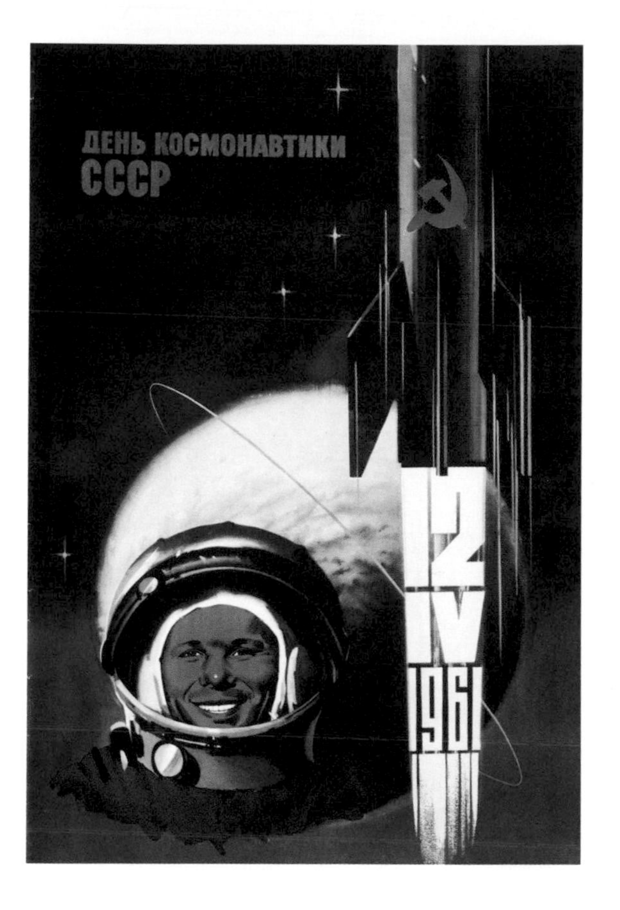
ДЕНЬ КОСМОНАВТИКИ
СССР
12
IV
1961

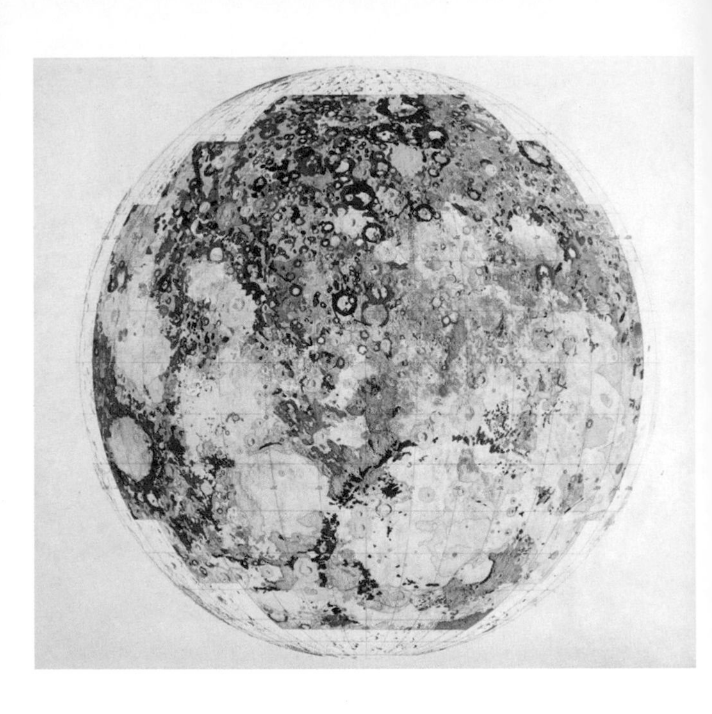

FEBRUARY

WEEK **08**

19
MONDAY

Holiday, USA (Presidents' Day)

20
TUESDAY

21
WEDNESDAY

22
THURSDAY

23
FRIDAY

24
SATURDAY

25
SUNDAY

Geological map of the moon, based largely on photographs taken by the US Lunar Orbiter 4 spacecraft. Launched on 4 May 1967, Lunar Orbiter 4 was the fourth in a series of five spacecraft designed to assist in the selection of landing sites on the Moon for the manned Apollo missions.

FEBRUARY • MARCH

WEEK **09**

26
MONDAY

27
TUESDAY

28
WEDNESDAY

01
THURSDAY

St David's Day

02
FRIDAY

○

03
SATURDAY

04
SUNDAY

Early X-ray of two goldfish and a sea fish from *Versuche Uber Photographie* by Dr Josef M. Eder and E. Valenta, published in Vienna in 1896. X-ray or 'new' photography caused a sensation when it was discovered by German scientist Wilhelm Conrad Roentgen in 1895.

Zwei Goldfische und ein Seefisch (Cristiceps argentatus).

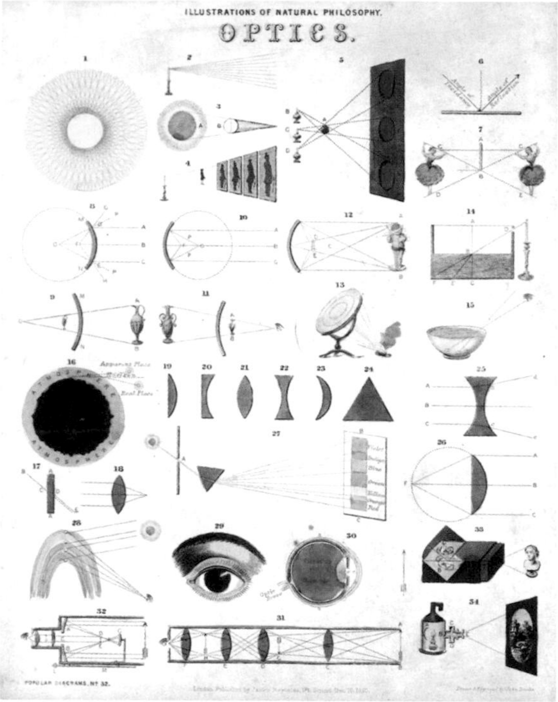
ILLUSTRATIONS OF NATURAL PHILOSOPHY.
OPTICS.
1
2
3
4
5
6
Angle of Incidence
Angle of Reflection
7
8
9
10
11
12
13
14
15
16
Apparent Place
Real Place
ATMOSPHERE
17
18
19
20
21
22
23
24
25
26
27
Violet
Indigo
Blue
Green
Yellow
Orange
Red
28
29
30
Optic Nerve
31
32
33
34
POPULAR DIAGRAMS, Nº 32.

MARCH

WEEK **10**

05
MONDAY

06
TUESDAY

07
WEDNESDAY

08
THURSDAY

09
FRIDAY

☾

10
SATURDAY

11
SUNDAY

Mother's Day, UK and Republic of Ireland

Hand-coloured engraved plate by John Emslie illustrating the science of optics, published in 1850.

MARCH

WEEK **11**

12
MONDAY

Commonwealth Day

13
TUESDAY

14
WEDNESDAY

15
THURSDAY

16
FRIDAY

17
SATURDAY

● St Patrick's Day

18
SUNDAY

The first reflecting telescope made by Sir Isaac Newton was shown to the Royal Society in 1668, and still resides there after more than three hundred years. This 1924 replica is by F.L. Agate.

May 10th Detached Cumulus.

α 10,500 ft β 11,700 ft

MARCH

WEEK **12**

19
MONDAY

Holiday, Northern Ireland and
Republic of Ireland (St Patrick's Day)

20
TUESDAY

Vernal Equinox (Spring begins)

21
WEDNESDAY

22
THURSDAY

23
FRIDAY

24
SATURDAY

☽

25
SUNDAY

Palm Sunday
British Summer Time begins

Detached cumulus cloud formations taken at Kew Observatory in 1887.

MARCH • APRIL

WEEK **13**

26
MONDAY

27
TUESDAY

28
WEDNESDAY

29
THURSDAY

Maundy Thursday

30
FRIDAY

Good Friday
Holiday, UK, Canada, Australia and New Zealand

31
SATURDAY

○ Holiday, Australia (Easter Saturday)
First day of Passover (Pesach)

01
SUNDAY

Easter Sunday

'Canoe Broaching to, in a Gale of Wind at Sunrise', 23 August 1821, by Edward Finden after a drawing by George Back, one of the members of John Franklin's expedition, from Franklin's *Narrative of a Journey to the Shores of the Polar Sea, in the years 1819, 20, 21, and 22*, published in 1823.

Drawn by Lieut. Back R.N.

Engraved by Edw. Finden

CANOE BROACHING TO, IN A GALE OF WIND AT SUNRISE.

Aug. 23. 1821.

Published March 1823, by John Murray, London.

Noria.

APRIL

WEEK **14**

02
MONDAY

Easter Monday
Holiday, UK (exc. Scotland), Republic of Ireland, Australia and New Zealand

03
TUESDAY

04
WEDNESDAY

05
THURSDAY

06
FRIDAY

07
SATURDAY

08
SUNDAY

☾

Chromolithograph by E. Wormser of the bucket waterwheel or noria, part of a series produced for the purpose of instructing the general public in the intricacies of mechanical engineering, published in Paris in 1856.

APRIL

WEEK **15**

09
MONDAY

10
TUESDAY

11
WEDNESDAY

12
THURSDAY

13
FRIDAY

14
SATURDAY

15
SUNDAY

'Comparative Magnitudes of the Planets' by John Emslie, one of a set of teaching cards published by James Reynolds & Sons, London, during the 1850–60s.

PHASES OF SATURN

COMPARATIVE MAGNITUDES
OF
THE PLANETS.
PHASES OF VENUS

SATURN
NEPTUNE
URANUS
RELATIVE DISTANCES OF THE PLANETS FROM THE SUN.

7.

8.

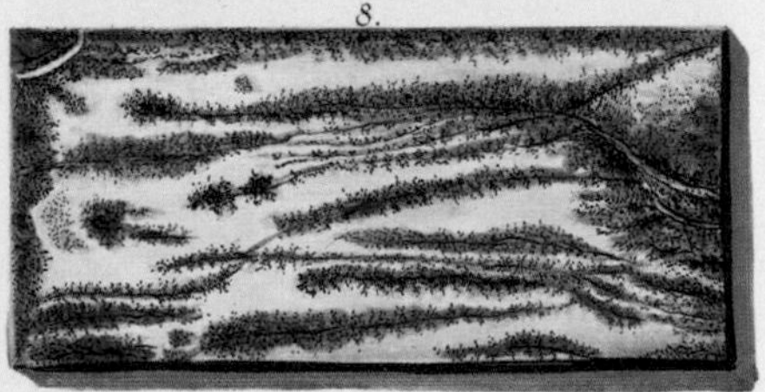

APRIL

WEEK **17**

23
MONDAY

St George's Day

24
TUESDAY

25
WEDNESDAY

Holiday, Australia and New Zealand (Anzac Day)

26
THURSDAY

27
FRIDAY

28
SATURDAY

29
SUNDAY

Guyton's 'Economical Laboratory', watercolour from 1797 showing an experimental set up. Baron Louis Bernard Guyton de Morveau was a French chemist.

APRIL

WEEK **16**

16
MONDAY

●

17
TUESDAY

18
WEDNESDAY

19
THURSDAY

20
FRIDAY

21
SATURDAY

Birthday of Queen Elizabeth II

22
SUNDAY

☽

Cross-section of stone with fossils, coloured engraving from *Marmora et adfines aliquos lapides* (Marble and some related stones) by Adam Ludwig Wirsing, published in Nuremberg in 1775.

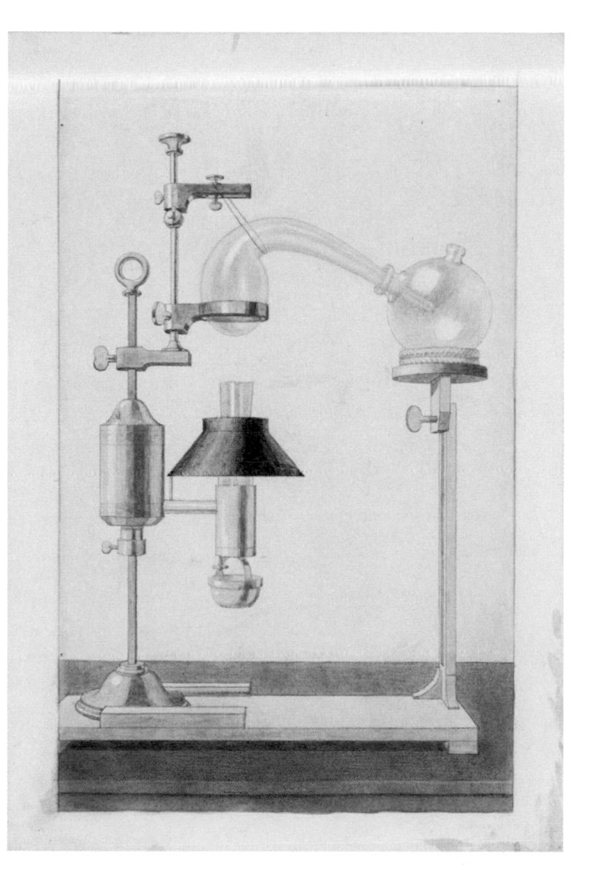

Burning Glasses, and Burroughs's Machine.

APRIL • MAY

WEEK **18**

30
MONDAY

01
TUESDAY

02
WEDNESDAY

03
THURSDAY

04
FRIDAY

05
SATURDAY

06
SUNDAY

'Burning Glasses, and Burroughs's Machine', 1799 book plate engraving by J. Pass, illustrating methods of creating fire by artificial means. The use of burning glasses was first noted in the writings of the Roman philosophers Seneca and Pliny the Elder, during the first century AD.

MAY

WEEK **19**

07
MONDAY

Early Spring Bank Holiday, UK
Holiday, Republic of Ireland

08
TUESDAY

☾

09
WEDNESDAY

10
THURSDAY

Ascension Day

11
FRIDAY

12
SATURDAY

13
SUNDAY

Mother's Day, USA, Canada, Australia and New Zealand

Landsat image of the Amazon, 1972. Five Landsats were launched between 1972 and 1984 to study the Earth's surface in various spectral bands from the visible to the infrared with a resolution of about 30 metres.

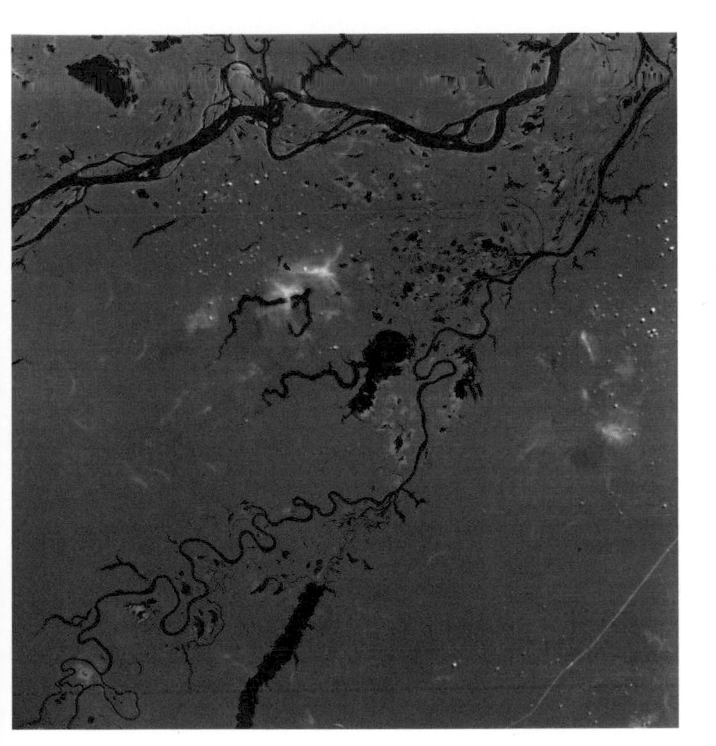

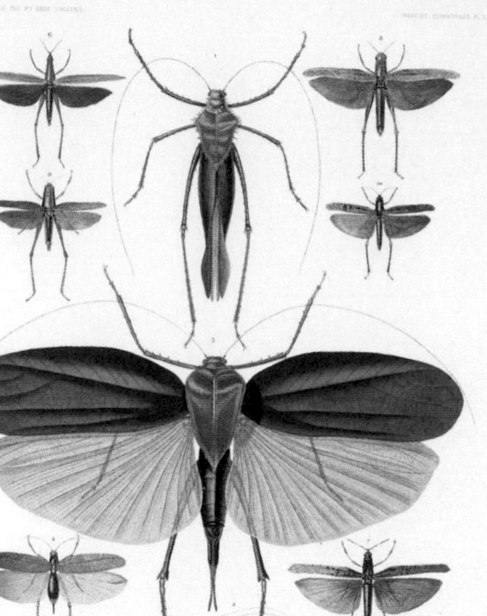

MAY

WEEK **20**

14
MONDAY

15
TUESDAY

16
WEDNESDAY

First day of Ramadân (subject to sighting of the moon)

17
THURSDAY

18
FRIDAY

19
SATURDAY

20
SUNDAY

Whit Sunday
Feast of Weeks (Shavuot)

Types of insect from Tenerife and Indonesia, an engraving taken from an account of a scientific expedition to study the geography, geology, anthropology and natural history of the South Pole and Oceania by French navigator Jules-Sebastien-Cesar Dumont d'Urville in 1837–1840.

MAY

WEEK **21**

21
MONDAY

Holiday, Canada (Victoria Day)

22
TUESDAY

☽

23
WEDNESDAY

24
THURSDAY

25
FRIDAY

26
SATURDAY

27
SUNDAY

Trinity Sunday

Educational poster by Emmanuel Dieuaide titled 'Tableau D'Aviation' (Aviation chart), depicting various flying machines designed between 1500 and 1880, arranged in a similar style to that favoured by butterfly and moth collectors. The publisher's own invention of 1879 is featured.

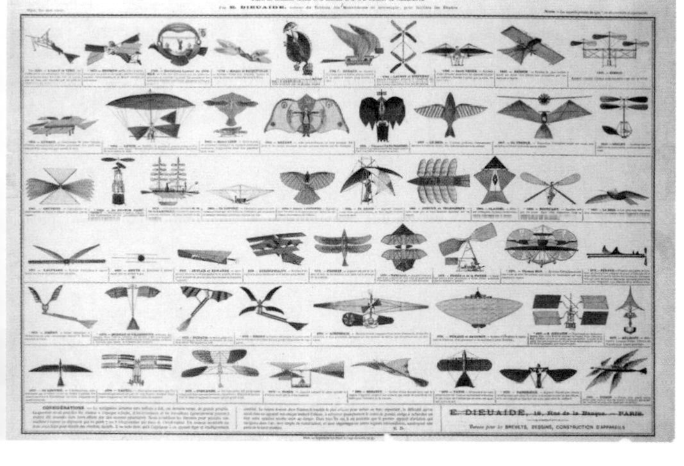
TABLEAU D'AVIATION
REPRÉSENTANT TOUT CE QUI A ÉTÉ FAIT DE REMARQUABLE SUR LA NAVIGATION AÉRIENNE SANS BALLONS
E. DIEUAIDE
PARIS
BREVETS, DESSINS, CONSTRUCTION D'APPAREILS

SHADOW OF THE RINGS ON THE PLANET
AT DIFFERENT SEASONS OF THE SATURNIAN YEAR.

MAY • JUNE

WEEK **22**

28
MONDAY

Spring Bank Holiday, UK
Holiday, USA (Memorial Day)

29
TUESDAY

○

30
WEDNESDAY

31
THURSDAY

Corpus Christi

01
FRIDAY

02
SATURDAY

Coronation Day

03
SUNDAY

Engraving illustration from *Saturn and its System* by Richard A. Proctor, published in 1865, showing the rings casting shadows on Saturn at different seasons of the Saturnian year.

JUNE

WEEK **23**

04
MONDAY

Holiday, Republic of Ireland
Holiday, New Zealand (The Queen's Birthday)

05
TUESDAY

06
WEDNESDAY

☾

07
THURSDAY

08
FRIDAY

09
SATURDAY

The Queen's Official Birthday (subject to confirmation)

10
SUNDAY

Hand-coloured engraving by Pierre-Joseph Buc'hoz of dendrites and agates, some of which have been enhanced by the artist to make them more remarkable, published in Paris between 1775 and 1781. At the time there was little understanding of the geological timescale of the formation of rocks.

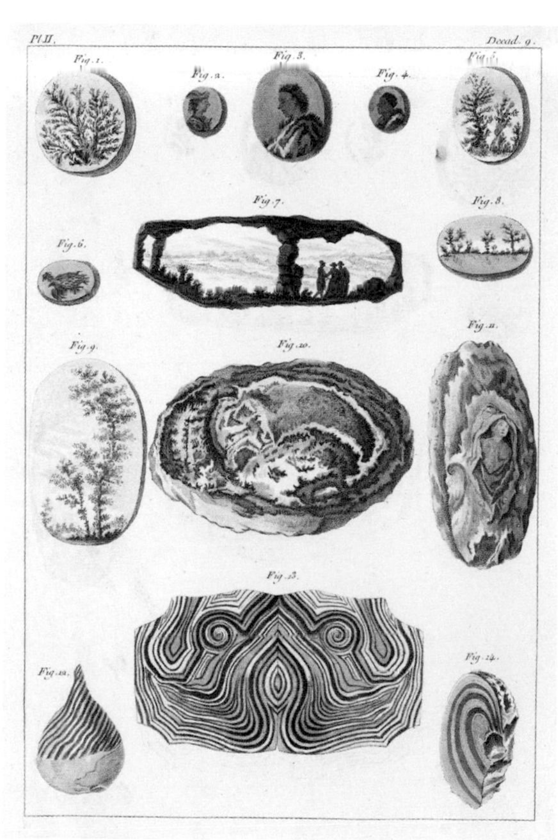

Pl.II.
Decad. 9.
Fig. 1.
Fig. 2.
Fig. 3.
Fig. 4.
Fig. 6.
Fig. 7.
Fig. 8.
Fig. 9.
Fig. 10.
Fig. 11.
Fig. 12.
Fig. 13.
Fig. 14.

AURORA BOREALIS

JUNE

WEEK **24**

11
MONDAY

Holiday, Australia (The Queen's Birthday)

12
TUESDAY

13
WEDNESDAY

14
THURSDAY

15
FRIDAY

Eid al-Fitr (end of Ramadān) (subject to sighting of the moon)

16
SATURDAY

17
SUNDAY

Father's Day, UK, Republic of Ireland, USA and Canada

Engraving of the Aurora Borealis observed at Bossekop in northern Norway on 19 January 1839, from *Electricity and Magnetism* by Amedee Guillemin, published in 1891.

JUNE

WEEK **25**

18
MONDAY

19
TUESDAY

20
WEDNESDAY

☽

21
THURSDAY

Summer Solstice (Summer begins)

22
FRIDAY

23
SATURDAY

24
SUNDAY

This case contains samples of five metals that were produced by electrolysis, dating from 1856–85. The metals are potassium, pure iron, strontium, calcium and lithium wire. Electrolysis was pioneered by the English scientist Humphry Davy.

JUNE • JULY

WEEK **26**

25
MONDAY

26
TUESDAY

27
WEDNESDAY

28
THURSDAY

○

29
FRIDAY

30
SATURDAY

01
SUNDAY

Canada Day

Detail of a view of the Earth as seen by the Apollo 17 crew travelling toward the Moon, December 1972.

JULY

WEEK **27**

02
MONDAY

Holiday, Canada (Canada Day)

03
TUESDAY

04
WEDNESDAY

Holiday, USA (Independence Day)

05
THURSDAY

06
FRIDAY

☾

07
SATURDAY

08
SUNDAY

Lithograph poster, 'Forward, the Komsomol Generation', by V. Yousov, Moscow, 1974.

ВПЕРЕД,
КОМСОМОЛЬСКОЕ
ПЛЕМЯ!

N *a given circle* 

to inſcribe an equilateral and equiangular pentagon.

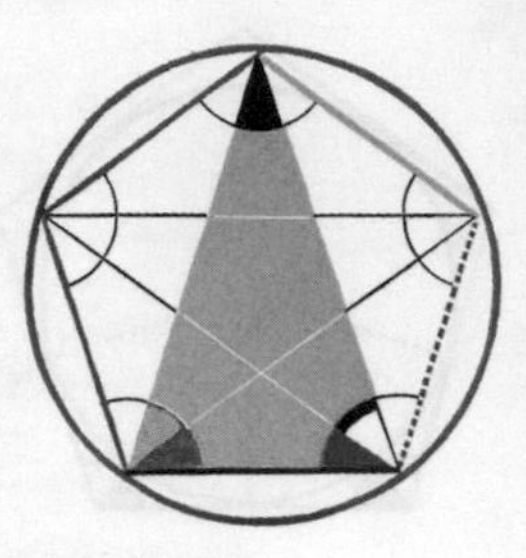

Conſtruct an iſoſceles triangle, in which each of the angles at the baſe ſhall be double of the angle at the vertex, and inſcribe in the given circle a triangle  equiangular to it; (B. 4. pr. 2.)

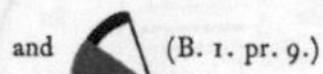

Biſect and (B. 1. pr. 9.)

draw , , and .

Becauſe each of the angles

 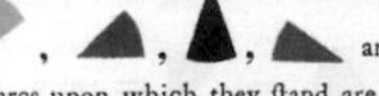 and 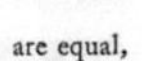 are equal,

the arcs upon which they ſtand are equal, (B. 3. pr. 26.) and ∴ , , , and which ſubtend theſe arcs are equal (B. 3. pr. 29.) and ∴ the pentagon is equilateral, it is alſo equiangular, as each of its angles ſtand upon equal arcs. (B. 3. pr. 27).

Q. E. D.

JULY

WEEK **28**

09
MONDAY

10
TUESDAY

11
WEDNESDAY

12
THURSDAY

Holiday, Northern Ireland (Battle of the Boyne)

13
FRIDAY

14
SATURDAY

15
SUNDAY

St Swithin's Day

Diagram of the method of drawing a pentagon within a circle using Euclid's theorems, from *The First Six Books of the Elements of Euclid: in which coloured diagrams and symbols are used instead of letters for the greater ease of learners* by Oliver Byrne, published in London in 1847.

JULY

WEEK **29**

16
MONDAY

17
TUESDAY

18
WEDNESDAY

19
THURSDAY

☽

20
FRIDAY

21
SATURDAY

22
SUNDAY

Engraving showing three sets of chemical apparatus: Machine for the Combustion of Phosphorous, the Mercurial Gazometer and Guyton's Eudiometer, from *Encyclopaedia Londinensis, or, Universal Dictionary of Arts, Sciences, and Literature*, published in London, 1810–1829.

CHEMISTRY. Plate VIII.

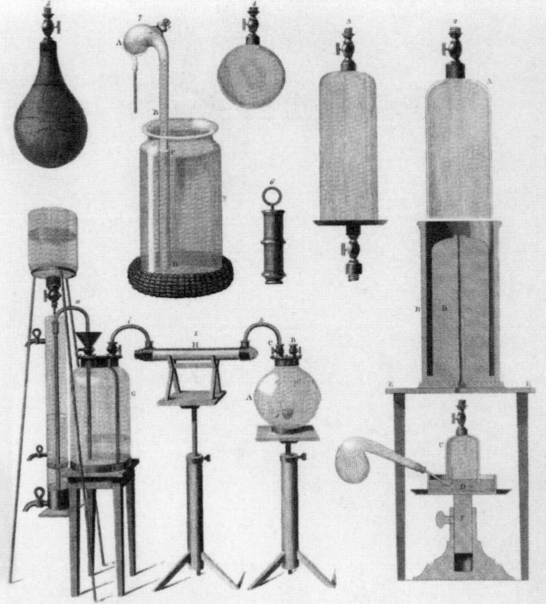

1. Machine for the Combustion of Phosphorus. 2 to 6. The Mercurial Gazometer. 7. Guyton's Eudiometer.

XXII ANDROMEDA
(the Chained Woman)
Pegasus
Al Phiratz
Pisces
Mirach
Cassiopeia
Al Maach or Al Amak
Perseus
Algol
Al Cenib
Caput Medusæ

JULY

WEEK **30**

23
MONDAY

24
TUESDAY

25
WEDNESDAY

26
THURSDAY

27
FRIDAY

28
SATURDAY

29
SUNDAY

The constellation of Andromeda (the Chained Woman) by Amy Manson, based on a drawing from *The Witness of the Stars* by E.W. Bullinger, London, 1895.

JULY • AUGUST

WEEK **31**

30
MONDAY

31
TUESDAY

01
WEDNESDAY

02
THURSDAY

03
FRIDAY

04
SATURDAY

05
SUNDAY

Engraving of Herschel's 20-foot telescope made in 1783, by J. Walker from a drawing by S. Bunce.

„VICTORI
5

AUGUST

WEEK **32**

06
MONDAY

Holiday, Scotland and Republic of Ireland

07
TUESDAY

08
WEDNESDAY

09
THURSDAY

10
FRIDAY

11
SATURDAY

12
SUNDAY

Postcard for a hot air balloon race over the French countryside, part of a collection of around 2,800 pictorial cards amassed by Winifred Penn-Gaskell from the 1890s to the 1940s, illustrating the development of international air post.

AUGUST

WEEK **33**

13
MONDAY

14
TUESDAY

15
WEDNESDAY

16
THURSDAY

17
FRIDAY

18
SATURDAY

☽

19
SUNDAY

Model E-1 Electric Build-it Set made by the Electric Games Company of Massachusetts, United States, c.1956–60. The set includes instructions for making working battery-operated toys, including a ray gun, traffic light, bike horn and telegraph key.

MAKE ELECTRIC TOYS AND GADGETS THAT REALLY WORK!
ELECTRIC BUILD-IT SET
Model E-1
• IT'S EASY • FUN
• AND EDUCATIONAL
• TEACHES PRINCIPLES OF ELECTRICITY
STOP
ELECTRIC BUILD-IT SET MODEL E-1
LAMPS
DOMES
DOMES
BATTERY CAP AND SPRING
BUZZER
INSTRUCTIONS
SCREW DRIVER
CODE KEY
WRENCH
BATTERY SIZE-D
BATTERY CASE

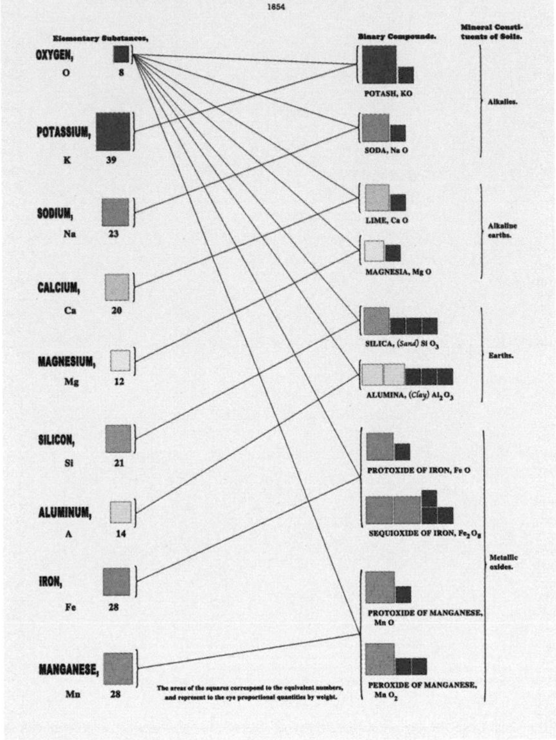

METALS AND THEIR COMPOUNDS WITH OXYGEN.
1854
Elementary Substances,
OXYGEN,
O 8
POTASSIUM,
K 39
SODIUM,
Na 23
CALCIUM,
Ca 20
MAGNESIUM,
Mg 12
SILICON,
Si 21
ALUMINUM,
A 14
IRON,
Fe 28
MANGANESE,
Mn 28
Binary Compounds.
POTASH, KO
SODA, Na O
LIME, Ca O
MAGNESIA, Mg O
SILICA, (Sand) Si O_3
ALUMINA, (Clay) Al_2O_3
PROTOXIDE OF IRON, Fe O
SEQUIOXIDE OF IRON, Fe_2O_3
PROTOXIDE OF MANGANESE, Mn O
PEROXIDE OF MANGANESE, Mn O_2
Mineral Constituents of Soils.
Alkalies.
Alkaline earths.
Earths.
Metallic oxides.
The areas of the squares correspond to the equivalent numbers, and represent to the eye proportional quantities by weight.

AUGUST

20
MONDAY

21
TUESDAY

22
WEDNESDAY

23
THURSDAY

24
FRIDAY

25
SATURDAY

26
SUNDAY

Illustration showing metals and their compounds with oxygen, 1854.

AUGUST • SEPTEMBER

WEEK **35**

27
MONDAY

Summer Bank Holiday, UK (exc. Scotland)

28
TUESDAY

29
WEDNESDAY

30
THURSDAY

31
FRIDAY

01
SATURDAY

02
SUNDAY

Father's Day, Australia and New Zealand

'Illustrations of Natural Philosophy: Electricity', drawn and engraved by John Emslie, published by James Reynolds, London, 1850, in response to the popular demand for information on the developments taking place in science and engineering as a result of the Industrial Revolution.

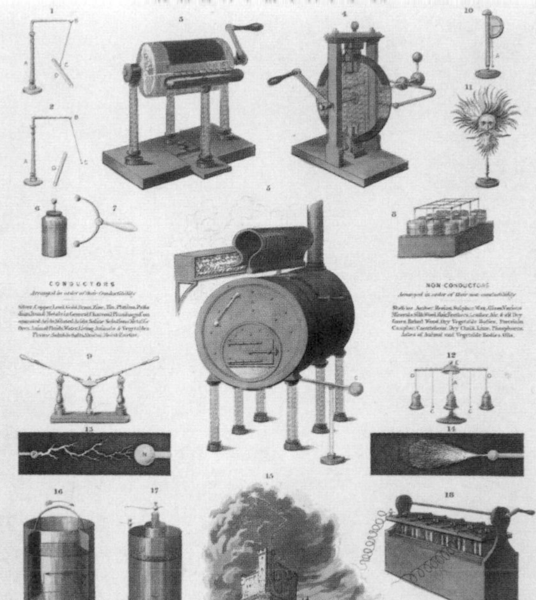
ILLUSTRATIONS OF NATURAL PHILOSOPHY.
ELECTRICITY.
CONDUCTORS
NON CONDUCTORS
POPULAR DIAGRAMS. Nº 33.

1905: 100 HP corsa
« Gordon Bennett »

1906: Brevetti Fiat - 4 cil.

1911: Modello 70 - 4 cil.

1911: 300 HP corsa - 4 cil

1913: Modello 4
4 cil. vert. ant. - 30/45 HP

1913: Modello 5
4 cil. vert. ant. - 50/60 HP

1922: 2 litri - 6 cil.
Corsa « G. P. di Francia »

1923: Modello 519

SEPTEMBER

WEEK **36**

03
MONDAY

☾ Holiday, USA (Labor Day)
Holiday, Canada (Labour Day)

04
TUESDAY

05
WEDNESDAY

06
THURSDAY

07
FRIDAY

08
SATURDAY

09
SUNDAY

●

Detail from poster showing 72 illustrations of Fiat car profiles from 1899 to 1954.

SEPTEMBER

WEEK **37**

10
MONDAY

Jewish New Year (Rosh Hashanah)

11
TUESDAY

12
WEDNESDAY

Islamic New Year

13
THURSDAY

14
FRIDAY

15
SATURDAY

16
SUNDAY

☽

'Terrestial and Lunar Volcanic Areas Compared', from the second edition of *The Moon Considered as a Planet, a World and a Satellite*, by James Nasmyth and James Carpenter, published in 1874.

PLATE VI

PORTION OF THE MOONS SURFACE.

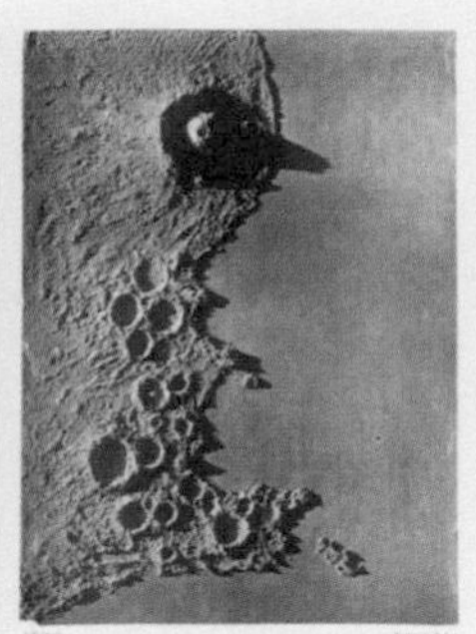

VESUVIUS AND NEIGHBOURHOOD OF NAPLES.

TERRESTRIAL AND LUNAR VOLCANIC AREAS COMPARED.

SCALE OF MILES.

Published by John Murray, Albemarle Street, Piccadilly.

telephonic offices the transmitter and receiver are sometimes joined by a bent spring, which serves as a handle, and they can thus be held simultaneously one to the ear, the other to the mouth.

Many other transmitters remain to be described, but as in most of these the fundamental part is a true "microphone," we will defer noticing them until we have given a description of this famous apparatus, the inventor of which is Professor Hughes.

Another class of battery transmitter is that in which a liquid is interposed at the working-point of the circuit. Elisha Gray devised

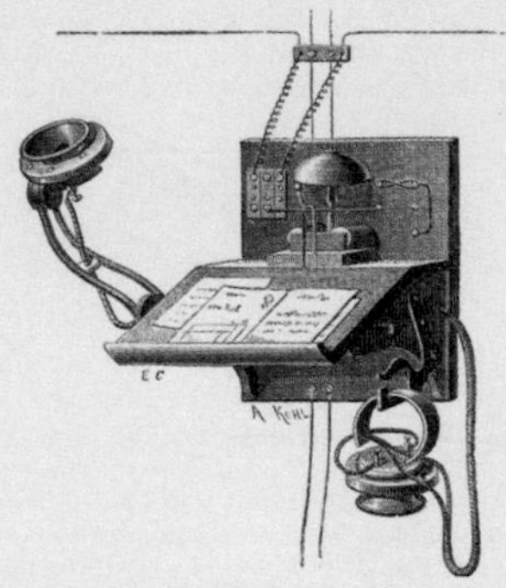

FIG. 445.—Telephonic set of instruments. (Edison's system.)

such a transmitter, in which a platinum wire attached to a diaphragm dipped down to a certain depth, nearly to the bottom, into a cup containing some conducting fluid, such as acidulated water. Other liquid transmitters have been devised by Graham Bell and by Edison.

§ 4.—*Telephonic Receivers of Various Forms.*

Since Graham Bell's researches, the arrangements of the receiver have been modified in many ways with the aim of improving it; most

SEPTEMBER

WEEK **38**

17
MONDAY

18
TUESDAY

19
WEDNESDAY

Day of Atonement (Yom Kippur)

20
THURSDAY

21
FRIDAY

22
SATURDAY

23
SUNDAY

Autumnal Equinox (Autumn begins)

Telephonic set of instruments (Edison's system) from *Electricity and Magnetism* by Amedee Guillemin, published in London in 1891. The American scientist and inventor Thomas Alva Edison redesigned the telephone shortly after its invention by Alexander Graham Bell in 1876.

SEPTEMBER

WEEK **39**

24
MONDAY

First day of Tabernacles (Succoth)

25
TUESDAY

○

26
WEDNESDAY

27
THURSDAY

28
FRIDAY

29
SATURDAY

Michaelmas Day

30
SUNDAY

'Illustrations of Natural Philosophy: Acoustics', drawn and engraved by John Emslie, published by James Reynolds, London, 1850.

ILLUSTRATIONS OF NATURAL PHILOSOPHY.

ACOUSTICS.

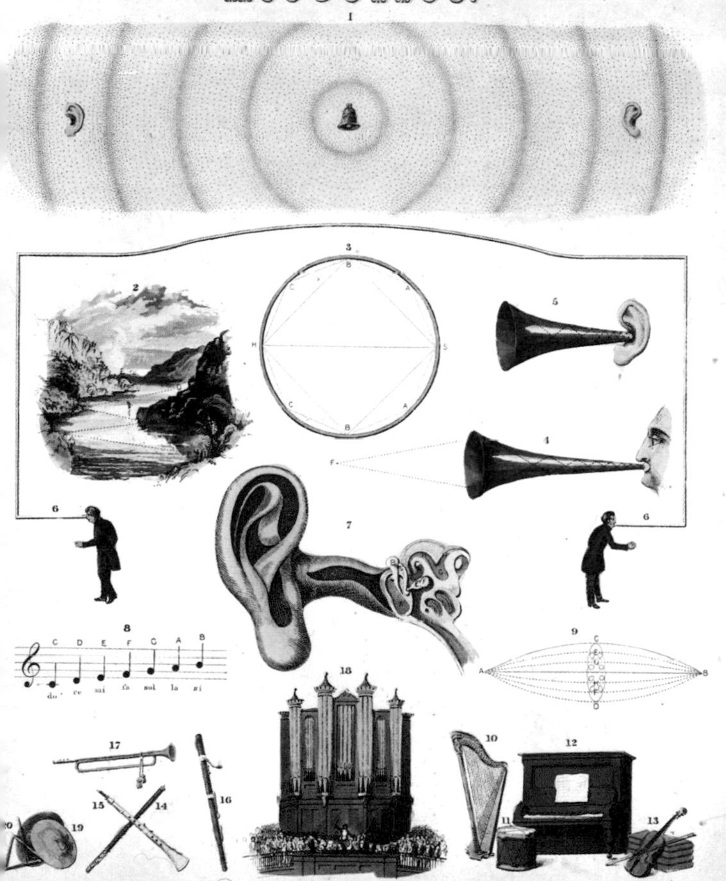

PULAR DIAGRAMS, No 31.

London Published by James Reynolds, 174 Strand Dec. 10 1850.

Drawn & Engraved by John Emslie

TAB : XXX.

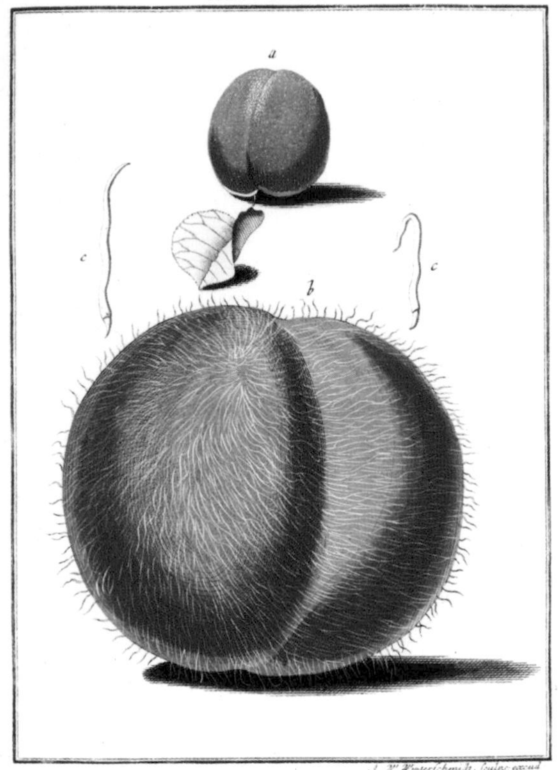

A. W. Winterschmidt, sculps. excud.

OCTOBER

WEEK **40**

01
MONDAY

02
TUESDAY

☾

03
WEDNESDAY

04
THURSDAY

05
FRIDAY

06
SATURDAY

07
SUNDAY

Peaches, 1776, plate from *Mikroskoopische Vermaaklykheden* ('Microscopic Delights of the Mind and Eye'), a Dutch edition of a book written by the German microscopist Martin Frobenius Ledermuller.

OCTOBER

WEEK **41**

08
MONDAY

Holiday, USA (Columbus Day)
Holiday, Canada (Thanksgiving)

09
TUESDAY

10
WEDNESDAY

11
THURSDAY

12
FRIDAY

13
SATURDAY

14
SUNDAY

Illustration of the lucernal microscope from *Essays on the microscope* by George Adams, published in London in 1787. Adams was instrument maker to King George III.

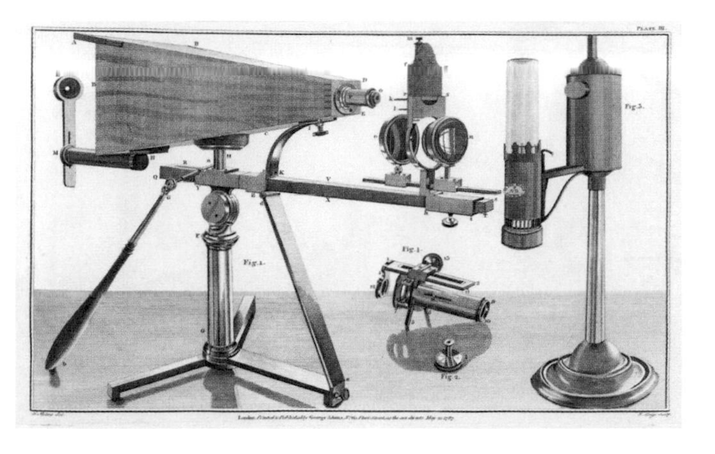

Emys vulgaris japon.

OCTOBER

WEEK **42**

15
MONDAY

16
TUESDAY

17
WEDNESDAY

18
THURSDAY

19
FRIDAY

20
SATURDAY

21
SUNDAY

Japanese terrapins by H. Schlegel, part of a work containing illustrations of new or lesser known amphibians, drawn from nature or from life, published in Dusseldorf, 1837–44.

OCTOBER

WEEK **43**

22
MONDAY

Holiday, New Zealand (Labour Day)

23
TUESDAY

24
WEDNESDAY

○

25
THURSDAY

26
FRIDAY

27
SATURDAY

28
SUNDAY

British Summer Time ends

One of fourteen x-ray positive prints illustrating the joining of the tibia and femur of a patient following the excision of his knee-joint, necessitated by injury sustained during the First World War, 1917–20.

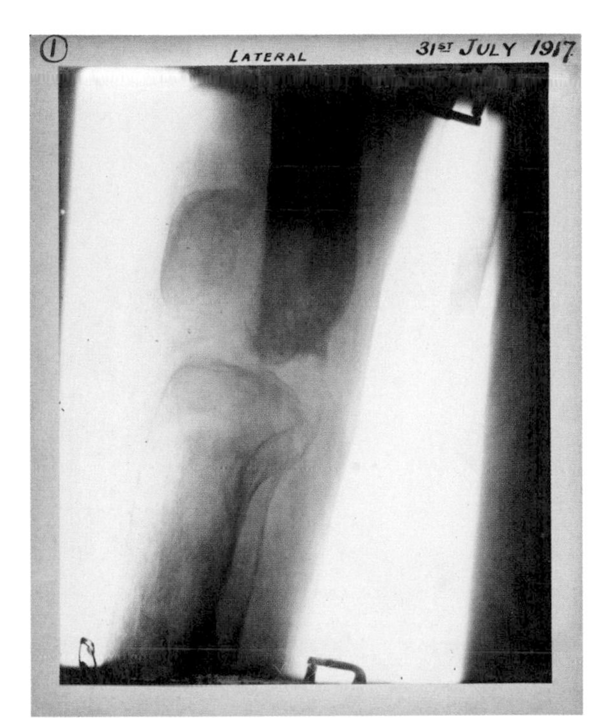
1
LATERAL
31ST JULY 1917

Plate 17.

GRAPHIC MICROSCOPY.

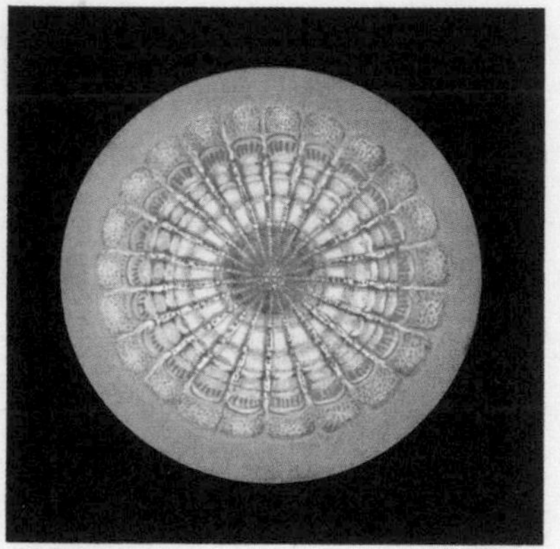

E.T.D. del, ad nat. Vincent Brooks, Day & Son lith.

TRANS. SECTION, SPINE OF SEA URCHIN (ECHINUS)

× 75.

"Science gossip". 1885.
Page 97.

OCTOBER • NOVEMBER

WEEK **44**

29
MONDAY

Holiday, Republic of Ireland

30
TUESDAY

31
WEDNESDAY

☾ Halloween

01
THURSDAY

All Saints' Day

02
FRIDAY

03
SATURDAY

04
SUNDAY

Chromolithograph illustration of a transverse section through the spine of a sea urchin, magnified by 75, dating from 1865–85.

NOVEMBER

WEEK **45**

05
MONDAY

Guy Fawkes

06
TUESDAY

07
WEDNESDAY

●

08
THURSDAY

09
FRIDAY

10
SATURDAY

11
SUNDAY

Remembrance Sunday

Photograph taken by Voyager 1 in 1979 showing a close up of the Great Red Spot on Jupiter, a storm that has been raging in the gas giant's atmosphere for at least three hundred years. The two Voyager spacecraft were launched in 1977 to explore the planets in the outer solar system.

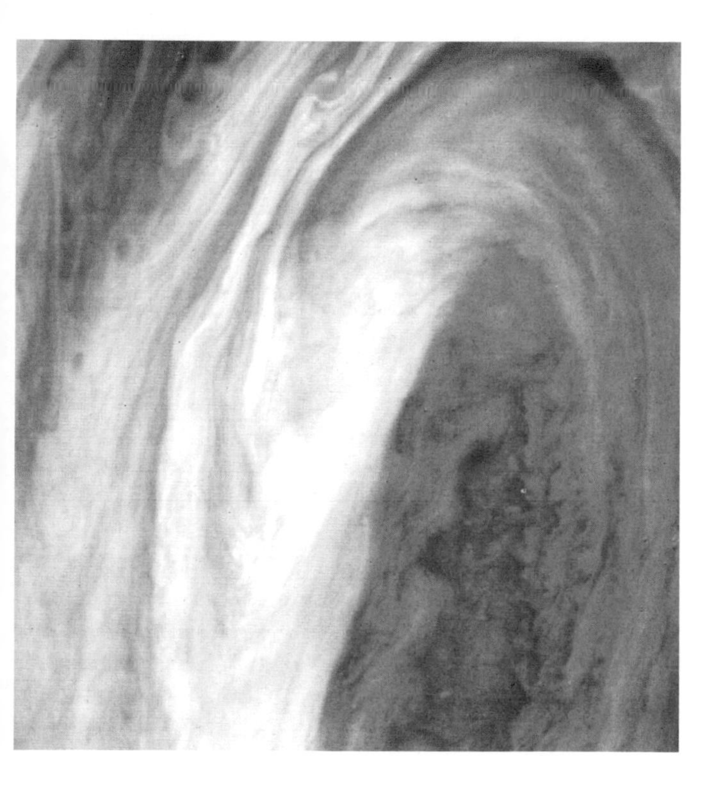

1771. - Nouveau MORANE-SAULNIER, monoplan Blindé
Type militaire blindé, l'avant est protégé par une cuirasse en tôle d'acier. Envergure 9 m. 30, longueur 6 m. 10
surface portante 15 mq. moteur Gnôme 80 HP.
J. H.

NOVEMBER

WEEK **46**

12
MONDAY

Holiday, USA (Veterans Day)
Holiday, Canada (Remembrance Day)

13
TUESDAY

14
WEDNESDAY

15
THURSDAY

☽

16
FRIDAY

17
SATURDAY

18
SUNDAY

Postcard for the new Morane-Saulnier Monoplane, published by Photo Edit of Paris, part of a collection amassed by Winifred Penn-Gaskell from the 1890s to the 1940s.

NOVEMBER

WEEK **47**

19
MONDAY

20
TUESDAY

21
WEDNESDAY

22
THURSDAY

Holiday, USA (Thanksgiving)

23
FRIDAY

○

24
SATURDAY

25
SUNDAY

Engraving showing 'Apparatus for the Formation of Phosphoric Acid by the Combustion of Phosphorus in Oxygen Gas' by Martinus van Marum, published in Haarlem in 1798.

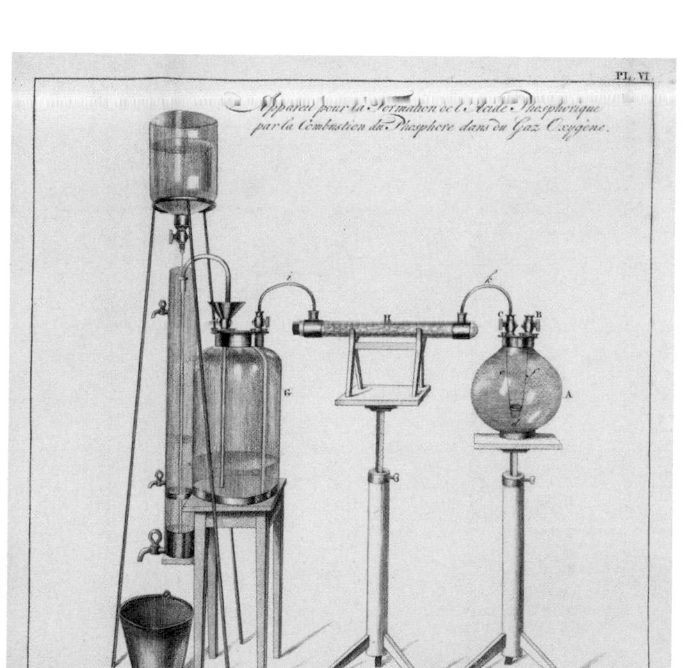
PL. VI.
Appareil pour la Formation de l'Acide Phosphorique
par la Combustion du Phosphore dans du Gaz Oxygène.
i
k
H
C
B
G
A

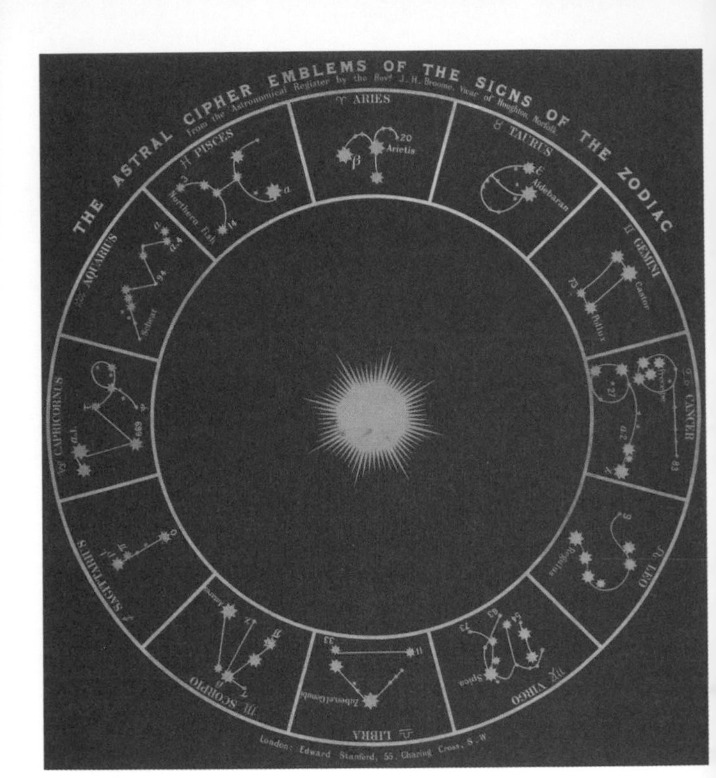
THE ASTRAL CIPHER EMBLEMS OF THE SIGNS OF THE ZODIAC
from the 'Astronomical Register' by the Revd J. H. Broome, vicar of Houghton, Norfolk.
♈ ARIES
♉ TAURUS
♊ GEMINI
♋ CANCER
♌ LEO
♍ VIRGO
♎ LIBRA
♏ SCORPIO
♐ SAGITTARIUS
♑ CAPRICORNUS
♒ AQUARIUS
♓ PISCES
Arietis
Aldebaran
Castor
Pollux
Regulus
Spica
Northern Fish
Scheat
London: Edward Stanford, 55, Charing Cross, S.W.

NOVEMBER • DECEMBER

WEEK **48**

26
MONDAY

27
TUESDAY

28
WEDNESDAY

29
THURSDAY

30
FRIDAY

☾ St Andrew's Day

01
SATURDAY

02
SUNDAY

First Sunday in Advent
Hannukah begins (at sunset)

'The astral cipher emblems of the signs of the zodiac' by the Reverend J.N. Broome, c.1870–80. This coloured print shows how the symbols representing the signs of the zodiac relate to the star patterns of the respective constellations in the heavens.

DECEMBER

WEEK **49**

03
MONDAY

04
TUESDAY

05
WEDNESDAY

06
THURSDAY

07
FRIDAY

08
SATURDAY

09
SUNDAY

Colour poster issued by the Soviet State Publishing House of Decorative/Fine Arts in Moscow to celebrate the first woman in space, Valentina Tereshkova, solo crew member on the Vostok 6 flight which launched from the Tyuratam Space Station on 16 June 1963 and remained in orbit for three days.

СССР
ВОСТОК-6
AŤ ŽIJE
PRVNÍ KOSMONAUTKA
SVĚTA!

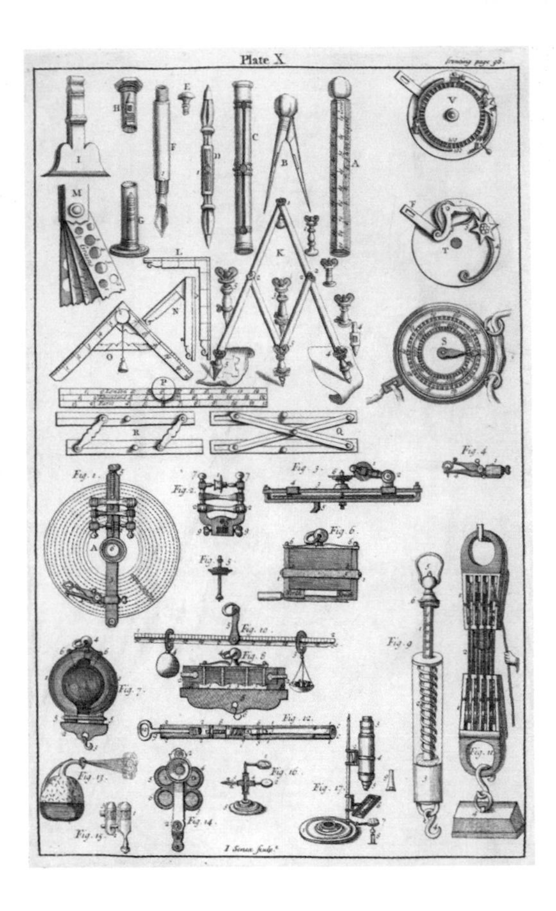
Plate X
fronting page 98.
J. Simon sculp.

DECEMBER

WEEK **50**

10
MONDAY

Hannukah ends

11
TUESDAY

12
WEDNESDAY

13
THURSDAY

14
FRIDAY

15
SATURDAY

☽

16
SUNDAY

Various instruments including a pantograph, scales and set square, from *The construction and principal uses of mathematical instruments* by N. Bion, published in London in 1723.

DECEMBER

WEEK **51**

17
MONDAY

18
TUESDAY

19
WEDNESDAY

20
THURSDAY

21
FRIDAY

Winter Solstice (Winter begins)

22
SATURDAY

○

23
SUNDAY

Plate showing Chichester Bell's 'Singing Water-jet', demonstrating that a jet of water is 'a machine for magnifying sound', taken from *Soap Bubbles* by Charles Vernon Boys, published in 1912. The 'Water telephone' was one of a great number of experiments devised by Chichester Bell.

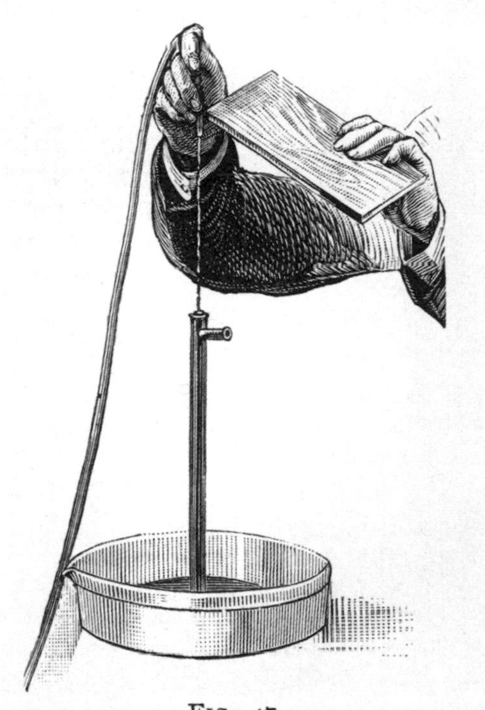

FIG. 47.

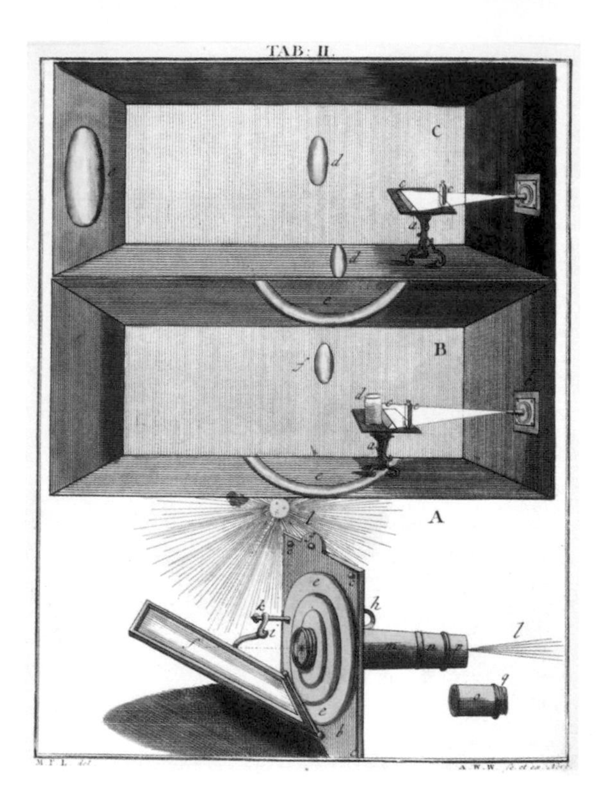
TAB: II.
C
B
A

DECEMBER

WEEK **52**

24
MONDAY

Christmas Eve

25
TUESDAY

Christmas Day
Holiday, UK, Republic of Ireland, USA, Canada, Australia and New Zealand

26
WEDNESDAY

Boxing Day (St Stephen's Day)
Holiday, UK, Republic of Ireland, Canada, Australia and New Zealand

27
THURSDAY

28
FRIDAY

29
SATURDAY

☾

30
SUNDAY

Solar microscope, 1776, plate from *Mikroskoopische Vermaaklykheden* ('Microscopic Delights of the Mind and Eye'), a Dutch edition of a book written by the German microscopist Martin Frobenius Ledermuller. The illustration shows how a solar microscope is used to produce a spectrum.

DECEMBER • JANUARY

WEEK **01 • 2019**

31
MONDAY

New Year's Eve

01
TUESDAY

New Year's Day
Holiday, UK, Republic of Ireland, USA, Canada, Australia and New Zealand

02
WEDNESDAY

Holiday, Scotland and New Zealand

03
THURSDAY

04
FRIDAY

05
SATURDAY

06
SUNDAY

● Epiphany

Thomas Edison's filament lamp with a single loop of carbon which glowed when a current flowed through it, made in 1879, the year after Joseph Swan's pioneering electric light bulb. Swan and Edison joined forces in 1883 to form the Edison and Swan United Electric Light Company.

NOTES